主编： 杨子江

绘画： 罗希贤　罗 一

编文： 沈嘉禄

Story behind Food

一菜一故事

上海書店 出版社
SHANGHAI BOOKSTORE PUBLISHING HOUSE

流金年代

中国是个公认的烹饪大国。烹饪大国地位的获得，首先取决于餐饮文化源远流长，积淀深厚。历代文化人对美食的关注或寄情，赋予食物以特殊的人文情怀，也体现了中华民族的价值观。

流传至今的与美食有关的传奇故事，一般都涉及重要人物，本意是借此表达中国人的道德情操，有惩恶扬善的劝喻作用。比如羊续悬鱼守廉、陶侃母亲封存腌鱼责问当官的儿子、羊祖延宴客一视同仁、陈遗孝母收集锅巴充饥、张翰生发莼鲈之思千里弃官、苏过烧制玉糁羹伺奉老父、苏轼被贬黄州煨制东坡肉、戚继光抗倭赶制光饼等等。反之，对石崇斗富办豪宴、张易之铁笼烤活鹅、王济创制人乳猪肉、孟昶不顾国之将亡大办船宴等"恶吃"表示不屑和讥讽。至于近现代传遍天下的名人美食故事就更加多了。

孔家花园创始人杨子江先生对美食故事的研究有很多年了。他对风云激荡的中国历史怀有浓厚兴趣，尤其是历史上的名人与美食的轶闻趣事，凡在书籍中读到便仔细记录在案。进入新世纪之际，他在徐家汇原孔祥熙别墅开了一家饭店，"就地取材"命名"孔家花园"。从此，他有了一个将美食故事进行实际演绎的"基地"。

不多久，孔家花园便以"一座有故事的花园，一家有好菜的饭店"名闻沪上，许多有故事的美味佳肴源源不断地端上餐桌，以飨食客，并经口口相传，由此成为美食家和各方宾客近悦远来的酬酢之处。孔家花园被有关方面评为"上海特色餐饮企业"、"上海最具餐饮文化中餐厅"。中央电视台《见证品牌》和《领航》两个节目组来上海采访，都特意找到杨子江先生，在《上海故事烩》和《品味民国菜》两档节目中用较大篇幅请他介绍通过美食来讲好中国故事的经验体会。

上海有四万多家饭店，为何央视独独选中孔家花园？就如央视导演所言：比好吃是没有底的，而餐饮能如此结合文化，至少目前，孔家花园是"独家"。因为导演和观众都想倾听菜肴中的传奇故事，这可以说是大型纪录片《舌尖上的中国》造成的边际效应。但更主要的原因应该是：好的中国故事本身具有的持久的说服力和感染力。甚至可以说，这为中国文化走出去提供了成功案例。

孔家花园老别墅店十年租约期满，完成了历史使命。但薪尽火传，春风再度，杨子江先生的美食"雕刻之旅"并未停止。他顺应市场要求，相继在浦东八佰伴商圈开出了孔家花园家宴馆，在五角场商圈开出了孔家花园民国主题餐厅，在淮海中路原华亭伊势丹开出了孔家花园霞飞店。杨子江先生对我说，他很想把已经离开"孔家"的"孔家花园"改名为"流金年代"，因为这些"有故事的菜"，都是中国历史上的吉光片羽，闪烁着中华民族的道德之光、智慧之光！

杨子江先生对经典名菜的"发现"和开发，代表了餐饮行业精英人士对饮食文化的热爱，对海派文化的深度研究，以及对餐饮风尚的精准把握。杨子江先生在菜肴设计上把握住了上海菜的"海派"精髓与拓展方向，通过拜访老一辈名厨，挖掘"有故事的美食"来满足人们物质与精神的双重需求。现在，杨子江先生主编了《一菜一故事》小册子，以人们喜闻乐见的连环画形式来讲述中国历史上的美食故事，由著名连环画家罗希贤先生、罗一女士执笔绘制，献给知味的美食家和连环画爱好者。书中收录的33道有故事的名菜由孔家花园（现拟改名为"流金年代"）经过研发筛选定型，日常供应，希望食家通过享用这些美食来感受中国文化的精妙之处和由此传导的正能量。

2016年5月

Preface

Executive Yang Zijiang of Kong's Garden restaurant has spent years digging into stories behind food. He reads a lot about the country's ups and downs. He's especially intrigued with the anecdotal history of food and the fame, many taking place in the *minguo*, the Republic of China (1912-1949). And he writes these curiosities down.

In 2001, Yang opened a Chinese cuisine restaurant in a European-style garden villa in Xuhui District, the former residence of Kung Hsiang-shi, the richest man in the early 20th century in China.

He named it Kong's Garden, after its former owner. Kung is spelled as Kong in Mandarin *pinyin*, the Chinese phonetic alphabet.

The menu features a selection of dishes, each of which is delivered with a back-story. Yes, it perfectly matches the nostalgic feeling that the lovely old house emanates.

Conceptualized by Yang as a fabulous restaurant nestled in a garden villa filled with stories, it soon became a hot dinning spot in the city, by word of mouth, from foodies to gourmets. And it's officially voted as one of most distinctive and most cultured local restaurants.

Later, Yang adopted franchising management and opened two chain outlets in commercial hubs in Pudong New Area and Yangpu District.

The fourth was opened in 2015 on the Middle Huaihai Road, one of the busiest roads in downtown over a century. Decorated in a beguiling *minguo* flavor, it awakens memories and recalls the past. When storytelling accompanies eating, it's a feast for stomach yet mind.

Recently, Yang was interviewed by China Central Television, which produced the popular Chinese food documentary "A Bite of China." He shared his experiences of how to use the food's narrative powers in two CCTV programs.

Shanghai has more than 40,000 restaurants. Why Kong's Garden?

"A Bite of China" has indeed, to some extent, triggered people's interests on yarns behind food. But the key reason, to me, is the communal enthusiasm for good stories, which are charming in nature.

It's also a great example of using food, a refreshingly accessible medium, to publicize Chinese culture.

Over years, Yang has visited many people, from skilled chefs, reputable gourmets to *lao ke la*, Shanghai dialect for the English word color or classic, which refers to elderly gentlemen born in high class and devoted to the Western good life of yore.

His efforts have paid. He heard exclusive stories and saw nearly-extinct recipes.

Yang epitomizes local elite catering people who make gastronomic pursuits a lifelong theme. They dive deep into *haipai* culture, literally Shanghai style, referring to the avant-garde but unique "East Meets West" culture of Shanghai.

Yang picked 33 dishes from his menu and illustrated in the "Story behind food" is a *lianhuanhua* comic book. Comic artist Luo Xixian painted for the book.

Diners are expected to feel our rich, positive and sophisticated culture.

Shen Jialu
May 2016

讲好中国故事
做好中国美食
传播饮食文化
彰显核心价值

Tell Chinese Stories, Make Chinese Dish.
Spread Food Culture, Manifest Core Value.

鱼腹藏羊

　　话说春秋时期，孔子带着大弟子子路游学列国，屡遭困厄，身心疲惫。一天黄昏，他们来到一条河边，前无村后无店，子路安排孔子先坐下休息，他四处去寻找食物。所幸这天运气不错，子路在半路上遇到一个猎人，肩负一只刚刚猎获的野羊，猎人就割下一条羊腿给他。子路往回走，又看到一个农夫从河里抓到几条鱼，农夫也给了他一条。

Yufu Cangyang (Roasted Fish Stuffed with Mutton)

　　Ancient Chinese sage Confucius and his disciple Zilu were said to have traveled through states during the Spring and Autumn Period (770-476 BC).

　　One dusk, they stopped by a river, finding they were in the middle of nowhere. The student was scouting around for food, when a hunter appeared. He cut off a leg from his newly-captured goat and gave to Zilu. On his way back for his master, Zilu ran into a fisherman, who sent him a fish.

　　New problem emerged when he returned to the riverside. He just picked bits of wood and sticks to make a fire, and he didn't have nothing to feed the pile. Obviously, the fire wouldn't last long, and it seemed that he had to choose which to roast and which to throw.

人物简介

　　孔子（公元前551年—公元前479年），名丘，字仲尼。春秋时期著名的大思想家、大教育家，儒家学派的创始人。孔子开创了私人讲学的风气，他的学说对中国与世界都有深远影响。孔子还被列为世界十大文化名人之首。

第一篇

　　天差不多快黑了,子路就赶快升起火来为老师做饭。但从河边捡拾的柴枝太少,烤了羊肉烤不了鱼。子路急中生智,将羊肉塞进鱼腹一起架在火上烤。不料才烤片刻,便奇香四溢,孔子闻香而起,品尝了一口便惊呼道:何物可以鲜美至此?鱼羊也!羊乃阳属乾,乾为天,鱼乃阴属坤,坤为地,鱼羊之合乃乾坤之合,鱼羊之鲜,乃天地所造!

　　古人认为鱼最鲜,所以最早的"鲜"字由三个"鱼"字组成。从此之后,"鲜"字改为用一个"鱼"字加一个"羊"字,"鱼""羊"为"鲜"就此开始。

　　Savoir-faire and sang-froid, he stuffed mutton into fish and roasted on fire. Confucius followed the delicious smell to Zilu's invention.
　　"Soooo umami!" he said, in an exclamation, after eating a morsel. He added that only the combination of fish and mutton could bring out such gastronomic delight.
　　Our ancients had appreciated the pure umami of fish most, and in Chinese characters, *xian* (umami 鲜), was initially composed of three *yu* (fish 鱼). After Confucius redefined the ultimate umami, *xian* was formed by *yu* on the left and *yang* (mutton 羊) on the right.

吃货说:
　　这道菜大有来头啊,绝对老古董级别。不知当年子路从渔民那里讨来的鱼是什么品种,但现在厨师选用的鲈鱼肉质鲜嫩,腹中塞进剁成末的羊肉,烧成之后吃不出一丝腥膻,味道真是一流。两种食材你中有我,我中有你,构成一场旷世绝恋,可见中国古人造字的机巧!

第一篇

带子上朝

孔子去世后,他的学说在西汉成为中华民族的主流价值观。孔子被尊奉为"大成至圣文宣王先师"。孔府子孙承袭一品显爵"衍圣公",则始于明代。清代乾隆皇帝将女儿下嫁孔府,以显示对儒学的尊重。乾隆皇帝还在特定节庆期间恩准孔府族长可以携带长子一起进宫朝圣,享受皇亲国戚同等待遇。

话说孔夫子第七十六代衍圣公孔令贻,带着儿子进京为西太后慈禧祝寿。孔夫人为迎接夫君和儿子回府,嘱咐孔府大厨烹制一道新菜,在洗尘家宴上给夫君一个惊喜。

Daizi Shangchao (Braised Duck and Pigeon in Soy Sauce)

Confucianism became mainstream values for Chinese society in the Western Han Dynasty (206 BC - AD 24), when a marquis title, Duke Yansheng, was created for his direct descendant.

The Duke Kong Lingyi, the 76th generation descendant of Confucius, had carried his son to celebrate birthday for the Empress Dowager Cixi in Beijing. To welcome them back, Kong's wife asked the family's private chef to create a new dish.

The chef thought it hard. Inspiration struck three days later.

He braised a duck, indicating Kong, and a pigeon, indicating his son, in soy sauce. He called it *daizi shangchao*, literally meaning father carried the son to the imperial court, just as the dish symbolized.

The Duke was greatly pleased. And it became a classic dish of Kong's family.

第二篇

带子上朝

孔府主厨苦思三日,终于想出用一只沂南麻鸭加一只潍坊野鸽制成一盘大菜,取名"带子上朝",以寄托"子承父业"、"光宗耀祖"的吉祥涵义,衍圣公品尝此菜后喜出望外。从此"带子上朝"就成了孔府家宴的经典名菜,每逢良辰佳节举办孔府大宴,必上"带子上朝"。后来在一鸭一鸽之外还加了一圈鸽蛋,显示孔氏家族的荣华富贵,世代相传。

吃货说:

当鸭子带上鸽子一起玩后,格局就大了。同为禽类,两种食材的细微差别还是让人分得出层次,肥腴、鲜嫩、清爽、细滑……比鸭子与鸽子单独成菜要丰富饱满得多。

霸王卸甲

楚汉之争，以项羽败亡、刘邦建立西汉王朝而告终。定都长安后刘邦举行开国大宴，犒劳有功之臣。张良献上一道大菜秘方，刘邦令御厨按秘方烹制，琴鼓声中红衣宫女端出大菜，只见金盘中央横卧着一只肥硕的大甲鱼，刘邦执箸品尝，只觉一种从未有过的鲜美绕舌而下，大喜之后遂问张良菜名，张良答道："此菜名为'霸王卸甲'，大鼋背甲好比西楚霸王项羽的盔甲，今日大王顺乎天意，卸其盔甲，天下归一，此时享用此菜，最合时宜。"刘邦听后哈哈大笑，挥手让群臣一起分享。从此，"霸王卸甲"成为汉代王侯庆功宴上的压阵大菜。

Bawang Xiejia (Simmered Turtles)

Following a peasant uprising that toppled the Qin Dynasty (221-206 BC), peasant leader Liu Bang of Han Power and aristocratic general Xiang Yu of Chu Power began fighting for the throne. The four-year convention ended with Liu's victory.

Liu treated meritorious followers a grand banquet. His strategist Zhang Liang offered the secret recipe of **bawang xiejia**, literally meaning shell being removed and indicating the failure of Xiang, dubbed bawang or dominator.

人物简介

刘邦与项羽

项羽（公元前232年—至前202年）与刘邦（公元前256年—前195年）各自起兵反秦。最后，刘邦灭亡秦朝，击败项羽，统一天下，成为汉朝开国皇帝，史称西汉。被誉为"其之神勇千古无二"的项羽，则只留下"霸王别姬"的千古绝唱。

第三篇

其实，张良献上这道名菜，也是为了提醒刘邦在改朝换代后尽快推行休养生息的政策。而他自己也不恋权位，辞让刘邦齐国三万户的赏赐，只在今天的江苏沛县请封留地，所以张良也称留侯。张良晚年又云游四方，修道养精，不知所终。

"It's the perfect time to enjoy this dish," Zhang said.

Liu chortled. He ordered the imperial chef to cook it and shared with his followers.

Bawang xiejia was then served as the last yet important dish at meals in honor of victories by Han imperial court.

吃货说：

家常一路的清炖甲鱼已不再让人兴奋，那么这道霸王卸甲以大气磅礴的体量和层次分明的味觉递进令人震撼。色泽金黄，汤汁浓稠，蒜香扑鼻，微辣上口，让真正的知味客领教了什么叫作具有复合味特征的秘制黄焖菜。

第三篇

小乔炖鸭

"遥想公瑾当年,小乔初嫁了,雄姿英发,羽扇纶巾,谈笑间,强虏灰飞烟灭 。"这是苏东坡在千古绝唱《赤壁怀古》中的咏叹,把那个怀拥美人,手握强兵,青年得志的白袍将军周瑜写得活龙活现,英姿飒爽,简直酷逼了。

Xiaoqiao Dunya (Stewed Duck with Eight Medicinal Herbs)

Famous military general Zhou Yu of the Three Kingdoms period (220–280 AD) had been sent by warlord Sun Ce to station troops in Sun's domain in Chaisang (present-day Jiujiang City of Jiangxi Province).

Day-to-day drills emaciated him, and his wife, Xiaoqiao, a famous beauty, made a duck soup for him. Eight medicinal herbs were stuffed in the duck. They included **dongchong xiacao**, a fungus that grows on certain plants.

It made Zhou feel rested and refreshed. After having the soup for a consecutive of ten days, he grew a ruddy complexion, and he felt full of vigor when making strategies and exercising soldiers.

人物简介

周瑜与小乔

周瑜(175年—210年),字公瑾,东汉末年东吴的统兵大将,著名美男子,足智多谋,文武全才。周瑜的夫人小乔与姐姐大乔均为绝世美女,分别嫁给周瑜和孙策。杜牧的名句"铜雀春深锁二乔",令这对美女穿越古今,青春长在。

小乔炖鸭

《烩金斋传》

话说东汉末年，东吴都督周瑜率军驻扎在柴桑（今九江市），他的爱妻小乔不但美丽，而且贤慧，看到夫君整日忙于指挥军队操演训练，日益消瘦，胃纳也差。于是想做一道菜给他滋补一下。她请教了名医后，选用当地肥鸭，在鸭腹内再加入冬虫夏草、泽兰等八味名贵药材，炖汤给周瑜吃。周瑜吃后顿觉神清气爽，浑身有劲，连吃十天，面容就变得红润光亮，排兵布阵更加有智有勇了。周瑜与刘备联手在赤壁大战中取得胜利，这份"爱的结晶"应该也有很大功劳吧！

吃货说：
　　一般人对药膳不感冒，但是这道名菜不尝一下的话，一对不起帅哥周瑜，二对不起美女小乔，三对不起踏遍青山尝百草的神农。再说这道炖鸭里加入的中药，味道还是相当温和的，不仅能滋阴养颜，补五脏之阴和虚痨之热，还有去腥、提香、增鲜等作用，一碗入肚，通体舒泰。

莼鲈之思

西晋文学家张翰的家乡在吴郡吴江（今苏州市吴江区）。张翰写得一手好文章，也是魏晋风骨的代表人物。齐王司马冏执政时期，征召张翰北上洛阳，封为大司马东曹掾。但不久张翰有感于政治污浊，社会混乱，决定退出权力中心，归隐山林。

Chunlu Zhisi (Sliced Perch Fish Stewed with Water Shield)

Litterateur Zhang Han of Western Jin Dynasty (265-316 AD) was born in Wujiang of Wujun (present-day Wujing District of Suzhou City).

Sima Jiong, imperial prince who briefly served as Emperor Hui's regent, noticed his writing talents and offered him an official position in the capital city of Luoyang.

On an autumn day, Zhang saw cool breeze swirling dusts, and he felt homesick.

"All of a sudden, wild rice stem, water shield and perch fish in my hometown came into my mind," Zhang told his friend.

人物简介

张翰，字季鹰，西晋文学家，江苏吴县人，一度在洛阳任职，创下了仅以"莼鲈之思"四个字留名于世的历史记录。

有一天，张翰看到一阵秋风卷起滚滚尘土，就跟身边的朋友说："我突然想到了故乡吴郡的菰菜、莼羹、鲈鱼脍，人生最重要的是能够适合自己的想法，怎么能够为了名利而跑到千里之外来当官呢？"于是他以"莼鲈之思"为名上书请辞，弃官还乡。

不久，齐王司马冏果然兵败，而有先知先觉的张翰幸免于难，后人将他视作具有独立人格精神的封建士大夫的典范。

张翰的诗文都已失传，只有他的"莼鲈之思"成了思乡的代名词，流传至今。

It occurred to him that what matters to a human being is to live the life they want to live.

He added, "How could I came so far just for fame and fortune?"

So, he resigned by reason of **chunlu zhisi**, literally missing water shield and perch fish.

吃货说：

　　太湖莼菜的嫩滑口感十分奇妙，含在嘴里有股清香充溢口腔，鲈鱼剔骨去皮披片，上浆余熟后，与莼菜一起构成一道风格典雅、口感清鲜的汤菜，一碗入口，遥想西晋时代的张翰先生，不禁心追神驰。

第五篇

洛阳燕菜

　　唐代武则天称帝时,有一个农民在洛阳东关荒地里挖出了一只重数十斤的大萝卜, 一时哄传,被当地官员当作祥瑞之物献给女皇。

　　武则天大喜,命御厨将这只前所未见的大萝卜做一道前所未有的萝卜菜。御厨苦思一夜,终于成功。

Luoyang Yancai (Shredded Turnip Nourishing Soup)

　　During the reign of the Empress Wu Zetian, a farmer dug out a big turnip in Luoyang, capital of the Tang dynasty (618–907 AD).
　　Rarely weighting dozens of pounds, it was seen as a good auspice and sent to the empress as a gift.
　　Under Wu's request, the imperial chef created a new turnip dish. He cut the turnip into shreds and steamed in nourishing soup, disguising as bird's nest soup when it was served to Her Majesty at lunch.
　　The empress didn't taste any sign of turnip from all 12 meals.

人物简介

　　武则天(624年—705年),名武曌,中国历史上唯一女皇帝,67岁即位,国号周。 82岁去世。

第六篇

第二天午宴之时，武则天无心品尝其他菜肴，专等前所未有的萝卜菜端上桌来。岂知十二道御膳上毕，仍未见萝卜菜上桌。武则天大怒，喝令御厨上殿，御厨见了女皇赶紧跪下禀告：陛下，第一道燕窝炖盅便是由萝卜所制！武则天不信，令御厨再上一道燕窝，细品之后，方觉此道燕窝真的与众不同，惊喜之下大赏御厨，并亲笔为此菜题了"假燕菜"三字。

后来此菜被列为唐代宫廷筵席第一菜，世称"洛阳燕菜"，又名"牡丹燕菜"，成了洛阳水席中的主菜。

She doubted the chef, and ordered him to make a bird's nest soup for her to compare. She savored it slowly, and finally, she tasted the difference.

Impressed and surprised, she sent the chef hefty reward and named the dish fake bird's nest, with her own inscription.

The dish was listed as the first course on the imperial banquet menu during the Dynasty Tang. People preferred to call it **Luoyang yancai**, literally a "bird's nest" dish in Luoyang. As it was presented like a blossomy peony, the dish was also called **mudan** (peony) **yan cai**.

吃货说：

想不到唐代就有叫厨师心惊肉跳的功夫菜啦！萝卜切细丝还不算太难，难的是入锅烩制后一丝不断，而且软糯适口，不柴不烂，吃在嘴里，没有萝卜的感觉，简直就是燕窝的升级版啊！清新爽口，醒酒妙品！

八宝豆腐

八宝豆腐是清朝康熙年间的宫廷名菜。

话说康熙皇帝第一次南巡来到苏州，住在织造府衙内。织造府主管曹寅殷勤伺侯，采办各种珍馐呈给皇上品尝，谁知康熙旅途劳顿，胃口极差，对各种山珍海味毫无兴致。这可急煞曹寅，他忽然想起姑苏名厨张东琯，立刻把他请到府上，重金相许，要他做出一道既清淡又鲜美，且富江南风味的菜肴，让康熙吃得开心。

张东琯不愧一代名厨，两天之后，一道佳肴便呈送康熙膳桌之上。康熙方品一口，顿觉胃口大开。原来这道菜肴以嫩豆腐为主料，辅以鸡肉末、猪肉末、火腿末、香菇末、蘑菇末、虾仁末、松仁末、瓜仁末等八味上好辅料，用鸡汤烩煮而成。

Babao Tofu (Eight Treasure Tofu)

Emperor Kangxi of the Qing Dynasty (1644-1911) had paid six visits to Jiangnan, region in the south of the lower reaches of Yangtze River.

人物简介

康熙皇帝（1654年—1722年），名爱新觉罗·玄烨，清朝第四位皇帝。8岁登基，14岁亲政，在位61年，是中国历史上在位时间最长的皇帝。他奠定了清朝兴盛的根基，被后世学者尊为"千古一帝"。

　　康熙龙颜大悦，将此菜命名为"八宝豆腐"，并将张东琯带到北京入职御厨房，赐五品顶戴，专烧八宝豆腐。康熙还命内臣将"八宝豆腐"的用料及烹调方法，写成方子藏入御膳房。每当大臣告老还乡，便将"八宝豆腐"配方相赐，作为奖赏。

　　不料大臣去领取配方时，给了御膳房大厨一个敲竹杠的机会，不奉上一千两银子的"红包"休想拿到，但"八宝豆腐"也就这样从宫廷流向民间。袁枚在《随园食单》中也记载了这个"取方费银一千两"的名菜。

He went to Suzhou in his first tour. Long journey drained him. But a freshly-made soup woke his stomach up. It was made by the renowned Suzhou chef Zhang Dongguan.

He stewed tofu and other eight ingredients, including minced pork, shrimp and mushroom, in chicken soup.

Kangxi liked it, and named it **babao doufu**, literally eight treasure tofu. And Zhang was promoted as an imperial chef.

吃货说：

　　作为烹饪大国，豆腐菜谱蔚为大观，无与伦比，但这道八宝豆腐应该是集大成者。豆腐洁白嫩滑，豆香浓郁清雅，名冠八鲜的辅料也非常到位，竭尽全力地将各自的本味释放出来，将一款豆腐提升到臻善臻美的境界。

乾隆鱼头

乾隆皇帝一生六巡江南，话说有一次他微服来到杭州吴山观赏美景，不料天降大雨，只好跑到山脚下一户农民家的屋檐下躲雨。屋内主人发现外面有人，马上开门将乾隆一行让进屋来。

这个农民名叫王小二，家中贫寒，心地善良，他见浑身湿透的乾隆又冷又饿，便将前一天留下的一只花鲢鱼鱼头再加一块豆腐，煮了一锅汤招待乾隆。乾隆执匙一尝顿觉异常鲜美，将一大碗鱼头豆腐全部吃光，还意犹未尽。饭后天晴，乾隆问过主人姓名，道谢后离去。

Qianlong Yutou (Tofu and Carp Head Soup)

Like his grandfather Emperor Kangxi, Emperor Qianlong also toured Jiangnan six times.

When he was in Hangzhou, he took time to the Wushan Mountain. A sudden downpour stopped him enjoying mountain view. He and his servants sheltered under eaves of a rural house on the foot of the mountain.

The owner, Wang Xiao'er, heard noises and invited them in. He made a soup, using tofu and variegated carp head, to warm Qianlong who got soaked through, and looked cold and hungry.

人物简介

乾隆皇帝，清朝第六位皇帝，自1736年至1795年在位六十年。是中国历史上长寿的皇帝。他与他祖父康熙主政的时代，史称"康乾盛世"。

第八篇

乾隆鱼头

乾隆回到京城，对王小二家吃过的鱼头豆腐念念不忘。若干年后，乾隆再次巡游江南来到杭州，派人到吴山脚下找到了王小二，王小二这才知道当年来家避雨并吃过一餐的那位大官人原来是当今圣上。乾隆问他过得如何，王小二对乾隆叹了一口气："王小二过年，一年不如一年啊！"乾隆说："你的鱼头豆腐不是做得很好吃吗？何不自己开一家饭馆呢？"说罢便赏了一笔银子给他，笑道："这是当年的饭钱！拿去做本钱开店吧。"

不久王小二就在吴山脚下开了一家王小二饭庄，乾隆得知后又亲笔题了一块匾额，上书"皇饭儿"三字，派人送到杭州。王小二将金匾挂在店铺大堂，于是满城轰动，从此生意天天红火。"皇饭儿"佳话也流传至今。

After the emperor returned Beijing, the soup came into his mind from time to time. When he visited Hangzhou again some years later, he gave Wang many silver ingots(the currency of ancient China).
He told Wang, "It's for your soup. You can use it to open a restaurant."
Wang did so, and his business was quite prosperous.

吃货说：

　　江南风味，西湖情调。豌豆粉皮与花鲢鱼头齐心协力，相互渗透，使寻常食材化身为经久不衰的经典名菜，与白汤略有不同的是，红汤似乎更加入味。

郑板桥甲鱼

　　扬州八怪的"领军人物"郑板桥，是乾隆元年进士，外放山东潍县知县，他关心民生，发展经济，勤政清廉，深受百姓拥戴。

　　郑板桥五十大寿之际，他本想不事声张，与家人在平静闲适中度过。不料生日那天清晨，县衙门外人声鼎沸，老仆探明情况后向郑大人禀报：四乡八邻的老百姓不约而同赶来为大人祝寿，而且每人都一手拎着一只大甲鱼，一手拎着一只老母鸡，恳请大人收下。

　　郑板桥听了很感动，推门一看，盛况果然，便连忙将大家请进县衙大堂，邀请大家一起享用这些母鸡甲鱼。

***Zheng Banqiao Jiayu* (Braised Turtle with Hen in Soy Sauce)**

　　Qing Dynasty painter Zheng Xie, better known as Zheng Banqiao, enjoyed high reputation and wide support when heading the Wei County, Shandong Province.

　　He preferred a simple quiet family gathering instead of a grand lavish banquet for his 50th birthday. But his people wanted to celebrate his big moment together.

人物简介

　　郑板桥(1693年—1765年)，康熙秀才，乾隆进士。扬州八怪的代表人物。一生只画兰、竹、石。一句"难得糊涂"，成为后世文人的自况。

衙中厨师面对这么多的母鸡与甲鱼，一时不知所措。郑板桥胸有成竹地说道：我教你用古法来焖制美味！于是面授机宜，厨师如法炮制。一个时辰后，只见一盘盘甲鱼焖鸡端了上来，色泽鲜黄，香味扑鼻。郑板桥向众乡亲敬酒频频，同声恭祝齐鲁大地物阜民丰，海晏河清。众乡亲则大快朵颐，争夸从未品尝过如此色佳味美的好菜，对郑大人平添了一份尊敬与爱戴！

鲁菜中极有特色的一道经典名菜"黄焖甲鱼"就此诞生，在民间也叫"郑板桥甲鱼"。

That morning, they gathered at his door, each with a hen and a turtle in hand. Both were highly-valued tonic.

They were so intense that Zheng accepted and asked them to come in for meal.

Zheng's private cooks were stumped by such an unexpected gift. To their surprise, Zheng had an recipe. As he said, cooks braised hen and turtle in soy sauce. Two hours later, it turned bright yellow and emitted a savory smell.

It then became a classic dish in Lu cuisine, or Shandong cuisine, one of the eight Chinese cuisine categories. Folks preferred to name it after Zheng Banqiao.

吃货说：

这是传统的黄焖烹法，但这道菜的焖制时间全凭厨师经验来掌控。甲鱼块与草鸡块相互帮衬，味浓而不厚，清鲜而不寡。滋味各参，香气夺人。

伊府面

　　伊秉绶任职惠州太守时,做了两件和他的生活与艺术有关的"大事":一是重修苏东坡故居时,意外从墨池里发现了苏东坡珍爱的"德有邻堂"端砚,据说这方端砚使他的书法作品充满了儒雅的气息。二是创制了流传至今的"伊府面"。

Yifu Noodle (Yi Mansion Noodle)

　　It's the story about maybe the world's first instant noodles.

　　Qing Dynasty calligrapher Yi Bingshou often ran gatherings when he was the magistrate of Huizhou City, Guangdong Province, making his home a popular salon where likeminded intellectuals rendezvoused for painting, poetry chanting, and of course drinking and eating.

　　It pleased every one except his private cooks who felt overwhelming to prepare food for so many guests in short time. Salon is widely regarded an engine for inspiration, and maybe because of that, Yi figured out a quick and easy recipe.

　　Cooks were told to dry and fry noodles. When guests visited, hey just need to add boiling water and toppings, and noodles were done.

　　Later, Yifu Noodle became a popular dish in Huizhou.

人物简介

　　伊秉绶(1754年—1815年),字祖似,号墨卿,福建汀州府宁化县人。乾隆五十四年进士,历任刑部主事、惠州太守和扬州知府。伊秉绶还是著名的书法家,尤其隶书,自成高古博大气象,也喜绘画治印,有诗集传世。

39

  话说伊秉绶在惠州的官邸，几乎就是士林云集的"文化沙龙"，志同道合的同僚与诗友经常聚在一起吟诗作画，一醉方休。家厨烧饭做菜有时忙不过来，伊秉绶便教厨师：用鸡蛋拌麦粉做成面条，入沸水锅中断生后捞起，卷成团，晾干后炸至金黄备用。客人来了，下水一煮回软，宜汤宜炒，加上荤素浇头，其味无穷。

  一次，诗人书法家宋湘到伊府作客，吃完这道美味的面后，问"此面何名?"知道尚未命名后便说："如此好面，岂可无名！不如就叫伊府面吧！"

  "伊府面"从此诞生，简称"伊面"，后来流传民间，成为惠州美食一种，并被后人称为世界上最早的方便面。

> **吃货说：**
>   伊府面出自名门，而且出自一位书法大家之手，故而它的基因里有文人墨客的浪漫情调和创新精神。海虾的鲜美汁液被面条徐徐吸入，面条的利爽口感又使得大虾特别鲜嫩滑爽。中国食材的配伍有时与中药相似，都是为了实现最终的和谐与升华。

红楼第一菜

　　古典名著《红楼梦》里涉及的美味佳肴有一百余道，但只有茄鲞的制作方法写得最为详实。所谓"鲞"，本意是剖开晾干的鱼，如"黄鱼鲞""笋鲞""咸鳓鲞"等，"茄鲞"就是用茄子做的茄子干。

　　这道名菜出现在《红楼梦》第四十一回，贾母叫王熙凤搛一些茄鲞给刘姥姥吃，凤姐儿依言搛些茄鲞送入刘姥姥口中，刘姥姥吃了不相信农民家里最寻常不过的茄子却有如此的味道："别哄我了，茄子跑出这个味儿来了，我们也不用种粮食，只种茄子了。"

Honglou No.1 Dish (Flavored Eggplant)

　　Hong Lou Meng, or A Dream of Red Mansions, is one of China's four great classical novels. It's a detailed record of life in the noble Jia Clan, vividly picturing their life of extravagance and exorbitance.

　　In chapter 41, Dowager Shi, the highest authority in the family, told her granddaughter-in-law Wang Xifeng to feed Granny Liu, Wang's distant relative from the country, flavored eggplant. It flabbergasted Liu who couldn't believed her tongue.

人物简介

　　曹雪芹(约1715年—约1763年)，古典名著《红楼梦》的作者。曹家三代为官，家世显赫，于雍正五年获罪抄家，从此败落。曹雪芹多才多艺，历经多年艰辛，终于写出了不朽巨作《红楼梦》八十回，可惜未完而逝。

第十一篇

于是凤姐儿告诉刘姥姥茄鲞的做法:"这也不难。你把才下来的茄子的皮签了,只要净肉,切成碎丁子,用鸡油炸了,再用鸡脯子肉并香菌,新笋,蘑菇,五香腐干,各色干果子,俱切成丁子,用鸡汤煨干,将香油一收,外加糟油一拌,盛在瓷罐子里封严,要吃时拿出来,用炒的鸡爪一拌就是。"刘姥姥听了,摇头吐舌说道:"我的佛祖!倒得十来只鸡来配他,怪道这个味儿!"

有红学家说,曹雪芹大概也就会做这道菜,在书中借着凤姐之口写出整个烹饪流程。从这个意义上说,此菜是不折不扣的红楼第一菜!

"It's not hard to make it," Wang said, and explained.
She added, "Eggplants are peeled, chopped and fried in chicken fat, and simmered in chicken soup, together with minced chicken breast, mushrooms, bamboo shoots, dried bean curd and mixed nuts. At last, the smorgasbord is seasoned with sesame oil and distillers' grains. It requires to be sealed tightly in porcelain."

The author Cao Xueqin detailed every step maybe because it was the only dish in the book that he knew the recipe. It could be considered as the No.1 dish in this regard.

吃货说:

跟谁混,决定了你的身价。土得掉渣的茄子从农民家的土灶头摇身来到贾府后,命运由此改观,鸡脯、火腿、笋尖、蘑菇等都成了它的配角。好在茄子还没有忘本,吸足了上等食材以及糟油的鲜香味后,还不忘初心。细细品尝,茄子还是茄子,只不过有着天下所有茄子都无法企及的风致。

第十一篇

老蚌怀珠

老蚌怀珠在清代有"江南一大名汤"之誉,这道汤品以河鱼为食材,再加蚌肉及腊肉、笋片、香菇等,用小火煨制一个时辰而成,汤清味鲜。而鱼腹中一定要塞入十几枚虾米猪肉圆子,故有"老蚌怀珠"的雅称,素为文人墨客看重。

曹雪芹虽然长住京城,却也会做一些江南好菜,而他最擅长做的便是"老蚌怀珠"。乾隆戊寅冬日,曹雪芹邀好友敦敏、于叔度来家中聚饮,于叔度上门作东,请曹雪芹以鱼为食材做一道菜。曹雪芹也不客气,立刻捋袖下厨,不一会便端上一只热气腾腾的海碗来。客人只见碗中卧着一条鳜鱼,鱼腹鼓鼓囊囊,一股鲜香之气扑鼻而来,拨开鱼肚,只见里面藏着一颗颗珍珠般的肉圆,迫不及待执匙品尝,并大加赞赏:诚为江南名菜,不枉上门作东!

Laobang Huaizhu (Stuffed Fish Soup)

It was dubbed as No.1 soup in Jiangnan during the Qing Dynasty.

Fish was stuffed more than a dozen small meatballs, made from pork and dried shrimps, and stewed with clam meat, preserved ham, bamboo shoots and mushrooms for two hours.

It tastes fresh and umami.

Intellectuals named it *laobang huaizhu*, literally pearls on clam, as an elegant metaphor to its appearance.

Though Cao grew up in Beijing, he was good at cooking Jingnan cuisine, and he was best at *laobang huaizhu*.

第十二篇

曹雪芹也非常高兴，频频举杯敬酒，三位好友在一片笑声中赋诗饮酒，度过了愉快的一天。曹雪芹的这段轶事被敦敏记入他的《瓶湖懋斋记盛》中，成了研究这位大文学家生平的重要内容。

吃货说：

不要以为这是上海石库门人家河鲫鱼塞肉的升级版，曹雪芹是大户人家出来的，曾经沧海难为水，什么没吃过啊，所以他传承下来的这道菜，自然有着不一样的诗性格调，是一次味觉与视觉的双重享受。当肉圆从鱼肚中滑出来时，仿佛听到林黛玉的那声惊呼！

第十二篇

红娘自配

话说晚清宫中有一个不成文的规矩,宫女年满二十岁后便要放她出宫嫁人。西太后慈禧身边有个贴身侍女名叫红喜,聪明伶俐,手脚勤快,善于鉴貌辨色,深得慈禧欢心。红喜一直到了23岁,却还不见太后有放她出宫的意思。红喜在家乡有一个青梅竹马的小伙子一直在等她回去成亲。花开花落,冬去春来,眼看马上就要变成"剩女",红喜心中不禁焦急异常。

一代名厨梁会亭是红喜的叔叔,当时正任职慈禧太后御厨房的主管。他看着侄女整天紧锁眉心,决定创造一道新菜,用这道菜去帮助侄女脱身。

Hongniang Zipei (Shrimp and Sea Cucumber in Soy Sauce)

It was an unwritten rule in the imperial court in late Qing Dynasty that palace maids could be freed when they reached the age of 20. But it didn't happen on Hongxi, one of the favorite attendants of the Empress Dowager Cixi, the supreme ruler at the time.

When Hongxi was 23 years old, she was still by the side of Cixi, serving her.

It worried her uncle Liang Huiting, the leading chef of the imperial kitchen. He decided to lend a hand to his niece.

人物简介

慈禧(1835年—1908年),即孝钦显皇后,中文名叶赫那拉氏。俗称西太后、老佛爷、慈禧太后。她是咸丰皇帝的妃子,同治皇帝的生母,先后立光绪、宣统为帝。因垂帘听政而成为晚清的实际掌权者。

于是有一天，慈宁宫午膳方始，一道新菜送到西太后面前。只见金盆之中，只只大虾面对条条海参，各具一色，似戏似舞，一看便知是一菜二吃。太后召来梁会亭问道："此菜何名？"梁会亭答:："红娘自配。"《西厢记》里大名鼎鼎的丫环红娘要自配郎君了！慈禧听了心中立即明白，却哼哼一笑::"原来虾要嫁与海参不成？一菜二味不错，只是菜名不佳！"

一月之后，红喜便出宫回乡，与心上人喜结连理。这道"红娘自配"也从此成了清宫名菜，成了相亲之前必先品尝的吉祥大菜。

One noon, he served a new dish to Cixi.

In a gold plate, fried shrimps and browned sea cucumbers were paired and placed face to face.

Cixi noticed it and ordered Liang to explain. "It's **hongniang zipei**," Liang said. Literally, it meant women want to get married. Cixi got his point.

One month later, Hongxi was freed the Forbidden City and married to her lover.

Hongniang zipei then earned fame in the Qing imperial court. Considered auspicious, it became a must-eat before blind dates.

> 吃货说：
>
> 　　一菜二吃，滋味复合，可以是兼融，也可以是对撞。此菜中的大虾参以西餐制法，蛋液上浆，温油炸酥，一口咬下，外脆里嫩，舌尖还能品出淡淡的芥茉味，开胃醒脑，回味甜鲜。海参则按照鲁菜"葱烤海参"之法烹制，卤汁紧包，软而不烂，适度弹牙。两道菜的口感味型与烹饪方法泾渭分明，忠实秉承了梁会亭的思路，还可能比当年慈禧所吃的更加美味。

第十三篇

栗子大葱

　　1900年，八国联军进攻北京，慈禧太后带着光绪皇帝"西狩"长安，途经山西泽州（今山西晋城）时，太后入住州府并吩咐进膳。但此时早过用膳时间，府衙厨房里只剩一堆山西特有的巴公大葱和一块猪肉，大厨怕食材太少不够慈禧一行食用，情急之下抓了一把去壳栗子扔进锅中同炒。等猪肉和大葱炒熟，他又怕菜中的栗子夹生，于是又将全部食材盛在碗里放入蒸屉去蒸，临时又往碗里撒了一把金钩海米提提鲜。旺火急蒸至食材全部酥烂，才抖抖索索送到慈禧太后与光绪皇帝面前。

Lizi Dacong (Chestnut Steamed with Scallion)

　　In 1900, the Eight-Power Allied Forces invaded China. Just before the besieged foreigners controlled Beijing, Cixi disguised and fled north to Shanxi Province, taking her nephew, Emperor Guangxu, with her.
　　One day, they arrived Zezhou of Shanxi en route to the provincial capital of Xi'an. They rested up from a helter-skelter escape in local government building, hoping to satisfy their hunger. But there were just some pork, scallions, and shelled chestnuts in the kitchen.
　　The chef stir-fried all he had and steamed them to make sure chestnuts were not raw. Before the dish was served, he sprinkled a handful of dried sea shrimps.
　　To his surprise, Cixi and Guangxu liked it.
　　Upon return to Beijing, Cixi ordered the chef to the palace to cook the dish for her again. The hurriedly-made dish, unexpectedly, became famous.

慈禧太后与光绪皇帝在北京宫中天天山珍海味，哪里有机会品尝到此等杂菜，而且此时又饿又乏，一尝之下，不觉胃口大开，对此菜的葱香、栗糯、肉鲜等特点留下深刻印象。

后来慈禧回到北京，特令晋城大厨上京再烹制此菜。消息传出，京城豪门大户争相仿效，弄得紫禁城内外葱香缭绕，食用栗子大葱成为一时风尚。这道"急就章"的菜肴，在意外中成了名菜。

吃货说：

　　栗子与大葱都是寻常食材，但产地不同，品质就有了高下之分，如果遇上高手在烹饪时的点化，会使它们获得价值提升。而且有里脊肉丝以及干贝的加盟，展开了一场热烈的"四角恋爱"，水火之间便形成了一种奇特的美味。当然，作为主角的大葱，每一根都忠实地传递出这份口感。

李鸿章杂烩

　　清光绪二十二年（1896年），李鸿章访问美国，在华盛顿中国使馆内宴请美国政要及各国大使。李中堂出国是带着各种中式食材和厨师团队的，于是他决定用一桌丰盛的中式豪宴来为大清帝国撑足面子。

　　且说那时候的英美洋人，大多没有来过中国，更不要说尝过中餐了。李中堂的这席中式豪宴，真正让洋大人见识了中国菜肴的五彩缤纷，尝到了中国滋味的变化无穷。在一片五体投地的惊叹声中，宴席告终，但意犹未尽的洋人们居然无一人肯起身离席，还在等待下一道菜。厨师长见状忙向李鸿章禀报：厨中备菜已尽，如何是好？李中堂听了微微一笑道：将桌上剩菜撤下，共煮一锅端来！

Li Hongzhang Zahui (Li Hongzhang's Hotchpotch)

　　In 1896, Qing Prime Minister Li Hongzhang held a banquet for US politicians and foreign ambassadors in the Chinese embassy in Washington.

　　The exquisite and sophisticated Chinese cuisine came such a delightful revelation. All dishes were served, but guests were still expecting to be surprised.

人物简介

　　李鸿章(1823年—1901年)，字渐甫，号少荃，安徽合肥人。晚清重臣，淮军与北洋水师的创始人和统帅，洋务运动的主要领导人。世人尊称他为"李中堂"，与俾斯麦、格兰特并称为"十九世纪世界三大伟人"。

第十五篇

片刻后只见一盆大菜端上餐桌,洋大人们认为收尾大菜必定隆重,一看其料丰富多彩,一品其味鲜美无比,于是纷纷请教李中堂此菜何名?这倒是李鸿章未经准备,于是连连说道:"好吃多吃,好吃多吃!"岂料"好吃多吃"正与英文"杂烩"(Hotchpotch)发音相近,从此"李鸿章杂烩"名闻天下;不单李鸿章后来宴请必备"杂烩"压阵,而且"墙外开花墙里红",从美国唐人街直至北京大栅栏,这道豪门杂烩一时红透天下。

As the kitchen ran out of ingredients, the chef turned to Li for help. Li told him to use what were left on the table and cook them together.

For a while, the impromptu stew was served. Of course, foreigners were amazed by diversified ingredients and umami taste. They wanted to know what it was, but Li just said "Haochi Duochi." literally meaning eating more if you like it.

It has the similar pronunciation with English word hotchpotch, and people stared to call it Li Hongzhang hotchpotch.

吃货说:

混搭有时会产生意想不到的奇效。这道汤菜中有许多上等食材,比如鱼翅、大虾、鱼肚、海参、山菌等,但画龙点睛的是一只由青笋制作的碧绿翡翠球和一只用蛋皮制作的黄色玉如意。这不仅是为了美观,还是一个暗示:这道看似庞杂的烩菜,其实有一段中西文化碰撞的外交佳话。

第十五篇

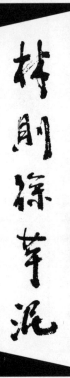

林则徐芋泥

福鼎八宝芋泥闻名天下,与清朝道光年间以禁烟而为世人敬仰的林则徐有关。

话说1839年,林则徐作为钦差大臣到广州禁烟,英美等国领事设宴接风,特备了西餐让林大人开开洋荤。林则徐应邀出席,当最后一道甜品上桌时,只见盘子里那砣白花花的东西冒着丝丝缕缕的"热气",林则徐便朝它吹了几口气,意在散热。谁知这道甜品是冰淇淋,它散发的其实是冷气,于是洋人相顾窃笑,将林则徐当作未见世面的乡巴佬。林则徐自知遭遇"暗算",但不动声色,只当无事。

Lin Zexu Yuni (Lin Zexu's Taro Paste)

In 1839, high-ranking official Lin Zexu was sent by Qing court to Guangzhou to suppress the opium trade that had devastated China's economy and poisoned the health of many Chinese in the 19th century.

Foreign consulates of US and Britain welcomed him with Western-style meal. The dessert was ice cream.

Lin had never seen it before. He blew on the "steaming hot" milky white thing as to cool it down, which triggered ironic snicker.

Some days later, it's Lin's turn to treat. He served them with **Fuding yuni**, taro paste from Fuding of Fujian Province.

人物简介

林则徐(1785年—1850年),字元抚,福建人。清代道光皇帝时官至一品,曾任湖广、陕甘、云贵总督,两次受命钦差大臣。史学界称他为近代中国"开眼看世界的第一人"。他坚决主张严禁鸦片,成为中国民族英雄。

第十六篇

　　过了几天，林则徐设宴"回敬"。筵席开始，几道凉菜之后，芋泥登场。洋大人一看此菜貌不惊人，颜色略带暗灰但又光滑透亮、表面没有一点热气，却喷香诱人，于是一个个举起汤匙，舀了满满一大匙送进嘴里。哪知这芋泥刚出锅，在猪油的掩盖下，热量全部积蓄内部，不明就里的洋大人大口吞进，一个个都烫得吐不了咽不下，眼泪直流，真正洋相百出。此时，林则徐悠然说道："各位大人，这是中国名菜槟榔芋泥，它外表冷静，内心炽热，就像我们中国人，与表面冒热气、里面冰冷的冰淇淋正好相反。"

　　林则徐话里有话地教训了这些外国使节，为中国人赢得了尊严。而惯于恃强凌弱的洋人们也真正领教了林大人的厉害。从此福鼎芋泥的另一个大名便是：林则徐芋泥。

Lard was used for a aromatic flavor and rich taste, and helped to trap heat in the taro paste. But the dim grey sheen was deceptive, disguising that it wouldn't hurt eaters.

Not realizing it, foreigners took a gulp. Very hot paste burned their mouths, scalded their tongues, and made their eyes water.

Lin said, "It's a well-known refreshment in China. Looking cold but tasting hot, it symbolizes Chinese people's characteristic, exactly contrary to your ice cream."

Since then, people called it Lin Zexu taro paste.

> 吃货说：
>
> 　　香浓馥郁、清甜滑润，加上多种果仁锦上添花，以及猪油的增香，使这道见证了清王朝外交史上餐桌风云的名点，有了老少咸宜的品质和宽容度。在微醺之后，由它来为一场欢宴收尾！

第十六篇

蒋公狮子头

蒋介石退居台湾后,经常去一位他的奉化籍老朋友奚炎先生家吃饭。奚老太太出身上海世家,做得一手好菜。蒋介石去奚家,叙旧之外,还能品尝奚家私房菜。

奚家私房菜以江浙菜为底子。奚老太太从菜市场采购普通食材,参以江浙菜的烹饪技法,做出一道道既有家常风味,又有筵席格调的菜肴,最经典的一道菜就是蛤蜊狮子头。

Jianggong Shizitou (Meatball in Clam)

After losing the civil war in 1949, Kuomintang leader Chiang Kai-shek and his forces retreated to Taiwan, where he established a government in exile.
Over years, he often went to dine at Xi Yan's home.
Xi was his old friend and hometown fellow from Fenghua in Zhejiang Province.
Xi's mother, born a well-known family in Shanghai, was a talented cook.
In young age, she learnt how to cook Jiangzhe cuisine, referred to cuisine from provinces of Jiangsu and Zhejiang, which features delicate flavors and emphasizes fresh ingredients.

人物简介

蒋介石(1887年—1975年),浙江奉化人。前中国国民党主席、总裁,国民党政府总统。

这道菜选用上好猪腿肉,斩成小粒后捏成一只橘子大小的肉圆,然后在蛤蜊鲜汤中小火煨数小时而成。关键一点要求肉圆在砂锅内不能松散,保持外形完整优美。蛤蜊的鲜味与瑶柱之类的干货鲜味有所不同,那是一种非常饱满的海鲜味,蛤蜊与猪肉两种鲜味互相渗透,完成了天作之合。

后来奚老太太将私房菜的技艺传授给了媳妇,为免蒋介石来往奚家的车马劳顿,还吩咐媳妇主动上门到士林官邸为蒋介石烹制私房菜,这款蛤蜊狮子头当然是必不可少的"保留节目"。蒋介石去世后,这位能干的奚家媳妇就写了一本书,记述了自己为蒋介石烧菜的故事,书名就叫《蒋公狮子头》。

Gradually, she started to cook up innovations at home, using simple and common ingredients. One of her signature dishes is *geli shizitou*, stewed meatball in clam.

When she was too old to cook, her daughter-in-law Yan Qiuli inherited her recipes, and cook for Chiang at his residence. Yes, *geli shizitou* was a must on the menu.

Yan preferred to call it Jianggong *shizitou*, literally Chiang's meatball. And she also used it to name her book, which revealed the family's secret recipes and Chiang's appetite. It was published after Chiang's death.

吃货说:

　　制作难点在于手工狮子头要在蛤蜊汤里久炖不散。肉圆要吸足蛤蜊的鲜,蛤蜊汤又要衬托猪肉的鲜美。水陆二鲜的联姻,在此成为天作之合。

第十七篇

蒋公豆腐

这道菜原名叫"奚家老豆腐",与"蒋公狮子头"所出同源,是奚家私房菜里的又一款经典菜式。因为蒋介石晚年几乎每餐必上此菜,侍卫们便称此菜为"蒋公豆腐"。

这道菜的制作要点,是要将整板豆腐焯水十余次,去尽石膏味,煮到豆腐呈现蜂窝状,将豆腐六面老皮切除,再将豆腐放入事先炖好的老鸡汤内,加入火腿丝、猪腿肉丝、鸡胸丝、冬菇丝、鲍鱼片、干贝、香菇等八种上佳食材,小火炖三小时左右。此时,原本十厘米见方的豆腐,缩到二厘米见方,一整板豆腐缩小到可为一只砂锅所容纳。豆腐的蜂窝孔内吸足了无与伦比的鲜汤,一入口中,鲜香嫩滑,诚为郁厨至味。

苏东坡曾发明东坡豆腐,治豆腐另有一功,他曾作诗称颂豆腐:"煮豆作乳脂为酥,高烧油烛斟蜜酒。"但苏东坡肯定没尝到过如此美味而富有营养的"蒋公豆腐"。

Jianggong tofu (Tofu in Chicken Soup)

It's another classic Xi's dish, which is tofu in chicken soup.

It's a must on the menu every time Chiang dined at Xi's home, and sometimes Xi's family delivered the dish to Chiang's residence.

In his old age, it became an essential dish in his every meal, and thus his guards called it *Jianggong doufu*.

蒋介石到奚家吃饭,每次不可无此菜。有时奚家也会做了这道豆腐送进士林官邸。有不知就里的人听说此事便说:老先生天天吃豆腐,真是俭朴得很啊!其实他不知道这块豆腐里包含了何种花样!

吃货说:

　　豆腐成为海绵,特制成蜂窝的豆腐,吸足了鲜汤。这是用八种上好原料炖出来的鲜汤,一口下去,鲜味在齿间弥漫开来……

黄埔蛋

据蒋介石的私人医生熊丸回忆,蒋介石早餐必不可少的一道菜是炒蛋。这不是一般的炒鸡蛋,而是蒋介石亲自命名的"黄埔蛋"。

1923年蒋介石接受孙中山委任,担任黄埔军校校长,他在军校里提倡吃"革命大锅饭",不准大吃大喝。但作为校长,蒋介石可以每天吃一个炒鸡蛋。毕竟整个黄埔军校只有校长有此待遇,蒋介石对这只炒鸡蛋相当看重。

黄埔蛋看似简单,其实很考验厨艺,曾经难倒过士林官邸内不少大厨。第一要保持鸡蛋的鲜嫩,要先加一定比例的水与料酒。第二须以人工打散鸡蛋,而且要达到将筷子插在蛋液中而筷子不倒的程度。第三是蛋液下锅后既煎又炒,获得外结壳里滑嫩的效果。

Huangpu Egg (Fried Egg)

Chiang couldn't have without Huangpu egg for breakfast. He grew fondness to it after he became the headmaster of the prestigious Huangpu Military Academy in 1923.

Chiang advocated populist and egalitarian. But he was given privilege, an additional egg every day.

Usually, egg was either scrambled or fried, but Huangpu egg was a daring mash-up, cooked in both ways.

Many Chiang's private cooks failed making it.

The Grand Hotel added Hungpu egg to its menu when Chiang's government treated foreign guests. Chiang had treated the US President Nixon with it.

由于蒋介石的偏爱，黄埔蛋后来被列入台湾圆山饭店招待外宾的菜单。尼克松访台时，蒋介石为了摆老资格，特别请尼克松品尝了这道黄埔蛋。

蒋介石患病时只能吃流汁，稍有好转可以进食了，他的要求便是：来一只黄埔蛋。

蒋介石晚年，宋美龄听从医生建议，禁止蒋介石食用胆固醇含量较高的黄埔蛋，蒋介石只好在招待客人时上这道菜，趁机一膏馋吻。

> 吃货说：
>
> 　　今天，有许多民间小菜都面临着失传。黄埔蛋也是濒临失传的广东传统老菜。蒋介石当年在广东偶然吃过一次，从此成瘾难忘。这道民间菜在蛋香之外，还散发着陈年佳酿的淡淡酒香，烹饪时既煎又炒，相当复杂。现在厨师配上宋美龄情有独钟的炸虾球，使蛋的色彩与口感趋于丰富。虾球用料新鲜，外脆里嫩，还有提香增鲜的芥末味。黄埔蛋配美龄虾球，这一菜式设计，为台湾圆山饭店所不及。

丰镐房芋头

"闯过三关六码头,吃过奉化芋艿头。"宁波人是一个具有开拓精神的族群,但他们无论走到哪里,都忘不了对家乡的奉化芋艿头发出如此的赞叹。

奉化芋艿头是宁波特产,外形粗糙,但肉粉无筋,质地细腻,入口软烂,具有滑、软、酥、糯的特点,制作菜肴时可以煨、烧、烩、烤,也可炒、拌、蒸、煮。芋头营养非同一般,不仅能调整人体酸碱平衡,还有防癌抗癌的功效。

Fenghao House Taro (Taro Simmered in Chicken Broth)

Ningbo people has a pioneering spirit, yet they are family people. Taro from Fenghua, Ningbo, is an unforgettable delicacy to them. So was for Chiang.

Especially, taro was so perfect for him in his old age when he had problems with teeth and stomach. His cooks invented many taro dishes and his favorite was taro simmered in chicken broth.

The dish was named after his birthplace Fenghao House.

丰镐房是蒋介石在家乡溪口的故居。丰镐二字取自西周的两个都城，即周文王的丰邑，周武王的镐京，为故居起名"丰镐房"，寄托了蒋介石建设"文武之家"的希望。

对奉化芋头的喜好伴随了蒋介石一生。对于牙齿与胃都有问题的蒋介石而言，芋头几乎是一种完美食材。厨师们为蒋介石创造了许多芋头菜肴，芋头肉丝、芋头白菜、芋头粉丝等等，但蒋介石最喜欢吃的是鸡汁芋头。

蒋介石官邸的厨师为蒋介石的"芋头系列"取了个名字："丰镐房芋头"。此中情怀，尽在不言之中。

吃货说：

芋头在奉化，有着别样的文化价值。此菜用鸡汤将芋头炖得很酥很酥，每一块芋头里都吸足了浓郁的鸡汤。这是蒋介石经常享用的家乡菜，如果加上虾仁等八宝浇头，那就锦上添花了。

爱庐蟹粉

1927年12月1日,蒋介石与宋美龄在上海举行了轰动中外的结婚大典。婚后,蒋介石夫妇长住南京,虽经常来沪,却无一适当住处。后来宋美龄的兄长宋子文物色到了今日东平路9号的一座花园洋房,买下作为妹妹的陪嫁,从此这幢房子成了蒋介石的上海行宫。

蒋宋二人在饮食口味上差异很大,霉豆腐臭冬瓜都是蒋介石的最爱,而宋美龄却偏爱木瓜牛奶等西方风味。但大闸蟹却是两人共同的喜爱。

Ailu Xiefen (Stir-fried Crab Roe in Orange)

Ailu, located at No.9 Dongping Road, is the former residence of Chiang and his wife Soong Mei-ling. The two-story villa was bought by Soong's brother Soong Tzu-wen and given to Chiang as a dowry when he married Soong's young sister on December 1, 1927. Chiang gave the villa a sweet name Ai Lu, literally love cottage.

Though the couple mainly lived in Nanjing, Ailu was one of their favorite residences.

Soong loved western food while Chiang had a purely Chinese stomach. But they shared the same fondness for hair crab.

人物简介

宋美龄(1897年—2003年),海南人。蒋介石夫人。她的大姐宋霭龄是孔祥熙夫人,二姐宋庆龄是孙中山夫人。宋家三姐妹以各自的经历,在中国近现代史上写下传奇篇章。

不过宋美龄对于大闸蟹的吃法，却是又有一功。宋美龄让厨师将炒好的蟹粉装在挖空的美国甜橙里，合上橙盖后再稍微加温，上桌后掀开橙盖，三根手指捏起橙盖轻轻一挤，橙汁的甜香与蟹粉的鲜美完成了"跨国联姻"。每年，在秋季的爱庐，宋美龄用这道自己独创的美食招待过无数来宾。

如今，爱庐已经成为上海音乐附中的校舍。"爱庐蟹粉"却流传至今。蒋介石亲笔题写的"爱庐"石碑仍在原处，默默注视着这世间的风雨变幻。

She figured out a novel way to enjoy crab in grace and elegance. She told cooks to stuff fried, crab roe, inside orange and then steam.

When it was served, diners lifted the top of the orange, which was cut, and squeezed juice to the golden crab roe.

In autumn when it's great to have hairy crab, Soong would treat her guests with *ailu xiefen*.

吃货说：

　　大闸蟹的鲜美，在美国脐橙汁液的"化学反应"下，一下子变得非常鲜美，而且没有一丝蟹腥味，回味有点甜，有点酸，馥郁芬芳，无与伦比。

第廿一篇

美龄赛熊掌

　　国民政府定都南京后,党政军机关云集总统府周边。1928年,有个叫李福全的人在总统府旁的丹凤街开了一家"双凤园"餐馆,专营京苏大菜,其中有一道极受欢迎的冷菜"赛熊掌"。

　　原来这道菜先把乌桕树的树叶捣烂制成汁,再将焖至半酥的猪蹄在黑色的树叶汁水里浸泡一天一夜,然后再上笼蒸至全熟,冷却后改刀装盆上桌。黑色的乌桕树叶汁中饱含花青素,这是极强的抗氧化剂,能够养颜美容,延年益寿。

Mei-ling Saixiongzhang *(Leaf Flavored Pig Feet)*

In 1927, Chiang moved his government to new capital of Nanjing.

　　Li Fuquan predicted huge demand and opened a restaurant, Shuangfengyuan, near the Presidential Palace in 1928. It served Beijing and Jiangsu cuisine. One of the most popular cold dishes is *saixiongzhang*, literally tastier to bear's paw.

　　It's actually made of pig feet. Leaves of Chinese tallow trees were crushed, and half-braised pig feet were infused with the dark tree sap. Twenty-four hours later, pig feet were steamed, and later cooled down.

　　Soong Mei-ling was very curious. She became a fan after tasting it.

　　南京市市长刘纪文、考试院院长戴季陶、监察院院长于右任、国民政府主席林森等,都是"双凤园"的常客。他们来店必点"赛熊掌",后来宋美龄听说这道奇特的菜后,也去品尝。可能是宋美龄知道黑色花青素对健康的益处,于是她成了"赛熊掌"最大的"粉丝"。因为这么多达官贵人喜欢"赛熊掌",这道菜便被当时南京人称为"黑官膳"。

　　1937年抗战全面爆发,南京沦陷,双凤园关门,"赛熊掌"这道名菜也湮没在历史长河中长达八十余年,直至如今方才"重出江湖"。

吃货说:

　　表面黝黑,不代表内心狂野,黑官膳也是如此。乌桕树叶的汁液给猪脚化妆,又使它的质地更加软韧可口,富有弹牙之感,还有一种天然植物的清香缓缓散发。品尝之间,似乎觉得眉宇间的肌肤神奇地滋润起来。

谭家大鱼头

谭延闿是一位颇有传奇色彩的民国名人,民国政界、文化界大名鼎鼎的美食家。谭延闿的贡献是提升了湘菜的品格。谭延闿祖籍湖南,长期任职广东又受到粤菜的影响,他对粤菜与湘菜的融合起了很大作用。湘菜的特点是汁厚味重,他将粤菜的清淡融入其中后,使湘菜的味道更加醇美,为湘菜日后跻身八大菜系打下了基础。因此谭延闿也被公认为"湘菜鼻祖"。

Tanjia Dayutou (Tan's Carp Head)

Tan Yankai was known as a gourmet in the political circle during the Republic of China (1912-1949). Born in Hunan Province, but working and living in Guangdong Province, he liked to mix and match two cuisines.

He refined a Hunan classic, fish head steamed with chopped chili. He used a secret sauce instead, giving a rich taste. The sauce was so umami that he often tossed it with rice or noodles.

It's one of his favorite dishes.

人物简介

谭延闿(1880年—1930年),字组庵,湖南人。曾任两广督军,三次出任湖南督军、省长兼湘军总司令。后任国民政府主席、行政院院长。有"近代颜书大家"之称。

谭家菜有"北谭""南谭"之说。"北谭"指清末官僚谭宗浚在北京所创的家传筵席。谭延闿的菜系又被称为"南谭"。两者并称清末民初两大官府菜。"北谭"后来家道败落，后人靠开饭馆谋生，名气虽因为食客增多而更大，菜品却不免流俗。而"南谭"则一直是私家菜，保持品位不降，一般人就难以吃到。

谭家大鱼头为"南谭"经典。取八斤重的花鲢鱼头一劈两半，加秘制酱料烹制，鱼头不烂不散，复合滋味层次丰富，不是一般的剁椒鱼头可相提并论。最经典的是其汤汁，秘制酱料的风味皆在其中，用来拌面拌饭，汤汁皆尽，食客方才尽兴！

吃货说：

　　秘制酱椒是这道菜的致胜法宝，如果厨师不明说的话，根本不知道其中还有干贝与火腿的加盟。湘菜与粤菜的组合拳，果然产生奇妙的感觉，吃完鱼头，还可以用鱼汤拌面，此时的面可能比鱼头更加好吃。

大千子鸡

国画大师张大千讲究生活品质,热爱美食,常常在艺术创作、游历各国的同时研究创新一些菜式,留下不少佳话。大千狮子头、大千豆腐、大千杂烩、大千鱼翅、大千鸡汁肝膏、大千牛肉面等,都是经他之手而流传于世的。这些创新美食最主要的特色在于"师古而不泥古"。张大千常跟朋友自夸:以艺术而论,我的烹饪水平应在绘画之上。

Daqian Ziji (Fried Chicken with Green and Red Peppers)

Zhang Daqian was a Chinese ink master, whose works could be auctioned thousands of millions of yuan. But he considered himself rather a good cook.

Zhang relished good eating. Recipes came to his mind from time to time whenever he was painting or travelling.

When he was in Japan, he taught a local Chinese cuisine cook how to fry chickens with green and red peppers.

The cook added it on his menu, and found it was so popular.

Zhang was also a big fan of it. Every time his friends visited him at home, he would cook it for them.

人物简介

张大千(1899年—1983年),字季爱,四川内江人。早期专心研习传统书画,后旅居海外,画风工写结合,重彩、水墨融为一体,尤其是泼墨与泼彩,开创了新的艺术风格,被誉为"当今最负盛名的国画大师"。

　　张大千在日本时，曾向东京的中国料理厨师传授了一道经他改良的新鲜菜式，这就是"大千子鸡"。这道菜以嫩仔鸡斩丁为主料，青红辣椒切丁为辅料，烹饪过程中又加入干椒、胡椒、花椒、郫县豆瓣、盐、糖、醋等调料，经急火爆炒而成，被后人誉为"大千风味"的顶峰之作。

　　在日本的中国厨师做成这道菜后，极受顾客喜爱。张大千也很喜爱自己发明的这道菜肴，每有好友来访，他必亲自捋袖下厨，一展手艺，饮酒吟诗，一尽雅兴。

吃货说：

　　此菜在口味上兼容了宫爆鸡丁的荔枝味、怪味鸡的怪味、豆瓣烧鸡的豆瓣味、糊辣鸡的糊辣味等，大千风味，名不虚传。张大千身前总有这么多美女围着他转，大概都是为了等吃他烧的美味佳肴吧。

爱玲鸭舌

作为才女作家和"小资教母"的张爱玲,从来没有放弃过对美食的追求,在她的作品中,无论小说还是随笔,写到美食总是充满激情和想象。在她吃过的美食中,对鸭舌的印象极其深刻。那是她童年时在天津吃过的一道鸭舌小萝卜汤,张爱玲在文章中的描述颇有情趣:"咬住鸭舌头根上的一只小扁骨头,往外一抽抽出来,像拔鞋拔……汤里的鸭舌头淡白色,非常清腴嫩滑。"

有位美食作家写道:"张爱玲所说'清腴嫩滑'的鸭舌,吃起来很有些像男女之间接吻的感觉,颇有销魂味道。张爱玲后来再吃鸭舌时,会不会想到她与胡兰成的拥吻呢?"

Zhang Ailing Yashe (Turnip and Duck Tongue Soup)

Renowned female writer Zhang Ailing, or Eileen Chang, lived a petit bourgeois lifestyle. She was so fastidious about food, and she has made gastronomic pursuits a revolving theme all her life. Her works showcased her fondness and enthusiasm for gastronomic delights.

人物简介

张爱玲(1920年—1995年),原籍河北唐山,生于上海。祖母是李鸿章的长女。1942年以小说《沉香屑·第一炉香》而在文坛一鸣惊人,代表作品有《传奇》、《倾城之恋》、《不了情》、《太太万岁》、《十八春》等小说、散文、电影剧本。1955年离开中国赴美定居。如傅雷所说,她的作品"是在一个低气压的时代,水土特别不相宜的地方开出的奇葩"。

  其实，中国的鸭舌菜有很多。清人所著的《调鼎集》中，就记录了不少鸭舌菜，比如糟鸭舌、白煮鸭舌、风鸭舌、煨鸭舌等。据《御香缥缈录》记载，慈禧太后非常喜爱的一道菜就是清炖鸭舌，上桌后必须安排在离太后最近的位置，便于她享用。慈禧与张爱玲有一样的爱好。

  张爱玲后来在上海以小说名世，直到她黯然神伤地离开大陆，这道令她销魂的鸭舌小萝卜汤，不知是否再次尝到？

  1995年，张爱玲躺在洛杉矶公寓的地板上，在这告别人间的最后时刻，望着窗外异国的天空，她会不会想起上海爱丁顿公寓阳台四周的万家灯火，还有那碗鸭舌汤？

One of the dishes that impressed her most was turnip and duck tongue soup, which she had in childhood in Tianjin.

"I bit on a small bone of the duck tongue and drew it out, feeling like I was pulling out the shoe horn," according to her article.

She continued, "The duck tongue was lightly white, very fresh, umami and smooth."

吃货说：

  小资教母不一定会做菜，但一定懂得如何吃。鸭舌在她的记忆中，与美味有关，更与情色体验有关。所以这道鸭舌的软滑、清鲜以及暗藏的那根软骨，均犹如一场初恋的跌宕起伏，值得久久回味！

第廿五篇

御香縹緲錄

德齡郡主

蜀腴牛四宝

电影《色戒》与张爱玲小说原著中,均以这样的开头叙述——几个贵夫人在打麻将,易太太说:"昨天我们到蜀腴去了,麦太太没去过。"结尾又与开头呼应:"还去蜀腴——昨天马太太没去。"可见蜀腴在当时上海人心中的至尊地位。

上海餐饮界是一个海纳百川的市场,早在上世纪二三十年代即形成了十八帮派聚齐的繁荣局面。十数家川菜饭店中,尤以蜀腴川菜社最受人青睐。

Shuyu Niusibao (Sichuan Flavored Beef Offal)

Shanghai is really a melting pot, and so does its catering circle.
Early in 1920 and 30s, China's main 18 cuisines were all accessible in Shanghai.
At the time, Shanghai had more than a dozen Sichuan restaurants, but people perferred to dine at Shuyu Chuancaishe, located around the corner of North Guangxi Road and Jiujiang Road.
Opened in 1937, it was a large restaurant, with five doors facing the street.
Sichuan cuisine master He Qikun was the one who helped Shuyu stand out. He refined classic Sichuan cuisine as to cater for local taste. One signature dish was tripe, ox toungue, tendon and veal in red oil, which was brightly red and slightly spicy.
Shuyu's dishes became so popular in Shanghai. Once, there was a trend to dine at Shuyu.

第廿六篇

蜀腴川菜社开设于1937年,地址在广西北路靠近九江路处,有五开间门面,在当时可算是一家规模较大的川菜馆了。开业之初,老板请川菜大师何其坤主理厨政,何其坤根据上海人的口味,对传统川菜进行了改良,在鲜香咸酸甜五味基础上,调和出七滋(麻、辣、咸、酸、甜、香、苦)八味(鱼香、麻辣、酸辣、干烧、辣子、红油、怪味、椒麻),令川菜的"百菜百味,一菜一格"登峰造极,一时间何其坤成为沪派川菜的领军人物。蜀腴的川菜风靡上海滩,各阶层人士、政府要员以及他们的太太小姐,到蜀腴去吃饭是一件很时尚的事情。以至数十年后,讲起蜀腴川菜,不但是老上海,在很多海外人士的心中,都会引起持久的怀念。

吃货说:

川菜素有七滋八味之说,这道菜则"入境随俗",成为海派川菜的经典名作。哇!食材表面泛起一层鲜亮的红油,入口后才发现仅仅是微辣而已,但鲜香无比,温和地刺激着味蕾与胃袋。出品成功的关键就是红油,据说是选用几十种调料秘制的。

第廿六篇

金必多浓汤

据台湾作家唐鲁孙撰文回忆，上世纪二十年代"上海南京路虞洽卿路口有一家晋隆饭店(今南京东路、西藏中路转角处)，虽然也是宁波厨师，跟一品香、大西洋，同属于中国式的西菜。可是他家头脑灵活，对于菜肴能够花样翻新，一只金必多浓汤，是拿鱼翅、鸡茸做的，上海独多前清的遗老遗少，旧式富商巨贾，吃这种西菜，当然比吃血淋淋的牛排对胃口"。

Jinbiduo Soup (Shark Fin and Chicken Soup)

Chinese author Tang Lusun, a distant relative to the Emperor Guangxu's favorite Zhen Concubine, once recalled his ultimate enjoyment in the Jinlong Restaurant at the corner of the East Nanjing Road and Middle Xizang Road dated back to 1920s.

"Like many others, it served Chinese-style western food. But its chef from Ningbo was quite smart and creative. He enlivened the menu and helped Jinlong to stand out," Tang said.

人物简介

唐鲁孙(1908年—1985年)，满族，他塔拉氏，镶红旗人，珍妃、瑾妃的堂侄孙。唐鲁孙出身贵胄，对老北京传统、风俗、掌故及宫廷秘闻了如指掌，年轻时游遍全国各地，见多识广，著有《中国吃》、《唐鲁孙谈吃》等，被誉为"华人谈吃第一人"。

第廿七篇

  1922年11月13日，爱因斯坦夫妇在福州路上的"一品香西菜馆"出席了为他举办的欢迎午餐会，上海黄浦区档案局保存的一份当年"一品香"套餐菜单，里面也记载着这道"鸡丝火腿鱼翅汤"。

  这道"奶油汤加一些儿鸡丝鱼翅"的中西合璧的汤品，正好聚集了中西菜品的要素，成为了海派菜"拿来主义"的经典作品。"金必多"的名字谐音"康白度"（买办），又响亮又吉利，一时风行十里洋场，成为当年中式大餐里的招牌菜。1949年以后，这道汤开始漂流到台北、香港，这道沉浮了一百多年的老汤，成为与百乐门爵士乐一样的老上海情怀的象征。

He added, "The chef added shark fin and minced chicken into cream to make an ultimate gastronomic delicate. It's more accessible to old fogies and young diehards who felt repelled by bloody steak."

There were allegations that *jinbiduo* was transliterated from the English's capital soup or comprador soup. Whatever, it made a hit in Shanghai, and became a signature dish in Jinlong and also other restaurants.

吃货说：

  十里洋场的西式浓汤，注入了中式原料：鱼翅与鲍鱼丝，迷失在它们陌生的奶油鸡汤里，却成就了一段名垂青史的中西联姻，为海派二字作出美妙的诠释。

第廿七篇

特卡琴科罗宋汤

　　上海淮海中路在上世纪二三十年代叫霞飞路,是当年外侨和华人心目中最繁华、最有情趣的一条商业街。霞飞路属于法租界,两万多名白俄登陆上海后,这里成为他们的避难与谋生之地,许多俄国商店在此开张,霞飞路变成了"东方的涅瓦大街"。霞飞路上有二十多家俄菜馆,比如文艺复兴、拜司饭店、DD'S、伏尔加、卡夫卡、库兹明花园餐厅等等,其中规模最大、环境最优的是特卡琴科兄弟咖啡馆,这是一家兼具咖啡馆和酒吧功能的俄菜馆,可同时容纳500多位客人,就餐时还有乐队演奏俄罗斯经典曲目。

Tkachenko's Luosongtang (Shanghai-style Borscht)

　　Today's Huaihai Road was called Avenue Joffre in 1920 and 30s, when it was the busiest and quaintest commercial street in Shanghai.

　　The area was held on lease by French imperialism. But gradually it was more like a Russian town.

　　After the Bolshevik Revolution in 1917, more than 20,000 White Russians fled to Shanghai, and many ended up in the French Concession because French was the second language of the Russian court.

　　Then, a rainbow of Russian shops and eateries were opened on the Avenue Joffre, which was dubbed as Oriental version of Neva Road of St Petersberg. They brought everything from Russia, from clothes to food.

　　Borscht is a classic Russian soup, but smart cooks, who followed Russians down south from Shandong, adjusted the recipe for locals. Shanghai didn't grow red beets, and they replaced tomatoes and ketchups.

　　As old Shanghainese called Russians *luosong*, the soup was called *luosontang*.

　　Tkachenko brother's Cafe was a big one among 20-plus Russian restaurants. And it served great *luosongtang*.

特卡琴科罗宋汤

潇洒金枪传

随白俄来上海为俄菜馆掌勺的厨师大多数是山东人，早年在哈尔滨俄租界学会了做俄式西菜。他们被上海人称之为"山东帮"，他们根据上海人的口味特点对传统俄罗斯菜进行改良，比如红菜汤，因为当时上海没有红菜头，他们就以番茄酱代替，很快适应了上海人的口味。因沪语把Russian读作"罗宋"，上海人把俄国人称为"罗宋人"，于是大名鼎鼎的"罗宋汤"就此诞生。

特卡琴科兄弟咖啡馆做的罗宋汤更为地道，成了"罗宋汤"的样板。

> 吃货说：
> 　　没有红菜头的上海如何做出浓郁红亮的红菜汤？这个难不倒俄菜馆里的中国大厨，他们在食材与调料上稍作改变，便成全了上海的食客，别具风味的罗宋汤由此成为上海流金岁月的不朽经典。番茄的微酸、胡萝卜的微甜、卷心菜的微软、洋葱的微辣、奶油的微浓，构成了一场跨越时空的旷世奇恋。

贵妃鸡翅

据说唐代杨贵妃平生最喜食两样东西，第一种是荔枝，第二种就是鸡翅。唐玄宗知道爱妃的这两种喜好，便令御厨反复研究，最后烹调出了一道荔枝味的鸡翅，令爱妃芳心大悦，吃得爱不释手。这便是川菜里的荔枝鸡翅。

杨贵妃晓音律，善歌舞。一日，唐玄宗约贵妃到百花亭赏花饮酒，数杯酒后，贵妃起舞助兴，此时的贵妃醉意朦胧，面似桃花，在明镜般的月光下，妩媚动人，美到无与伦比。这就是"杨贵妃酒醉百花亭"的故事，京剧大师梅兰芳的不朽之作《贵妃醉酒》便是根据这个故事改编的。

Guifei Jichi (Sweet and Sour Chicken Wings in Wine)

Yang Yuhuan, one of the top four beauties in ancient China, was the beloved consort of Emperor Xuanzong of Tang Dynasty during his later years.

The emperor knew lychee and chicken wings were Yang's favorite flavors, and ordered the imperial chef to invent a sweet and sour chicken wings. Of course, it greatly pleased Yang.

人物简介

梅兰芳（1894年—1961年），江苏泰州人，久居北京，中国京剧表演艺术大师，舞台生涯长达50年，开创了自己的艺术流派，代表作有《贵妃醉酒》、《天女散花》、《宇宙锋》、《打渔杀家》等。

第廿九篇

上世纪三十年代，梅兰芳在上海"共舞台"演出《贵妃醉酒》，全城轰动，万人空巷，一票难求。共舞台附近的一家川菜馆的大厨灵机一动，将他家所做的荔枝鸡翅加入上好红酒，烹出的鸡翅色泽金红，软滑细嫩，酒香浓醇，取名为"贵妃鸡翅"。声名传出，一时间顾客盈门，"贵妃鸡翅"与梅老板的《贵妃醉酒》同时誉满申城。

About 1,200 years ago, the recipe was refined by a cook in Shanghai.

At the time, Peking Opera master Mei Lanfang was impersonating Yang in the classic, Guifei Zuijiu, or Drunken Concubine, in the Shanghai Gong Stage. The chef was inspired to add wine, and named it **guifei jichi**.

Tender and mellow, it soon won popularity.

吃货说：

　　川菜的荔枝味，注入了老欧洲的葡萄酒，使平淡无奇的鸡翅一跃而成为一道具有异国情调的佳肴，酒香浓郁，荔枝鲜甜，让"贵妃"唱红上海大码头的时代背景也有了丰富的内涵。这又是一个海派经典！

第廿九篇 贵妃鸡翅

弄堂桂花肉

有一天，刚刚开门的孔家花园酒家急匆匆地走进一位中年男子，他问服务员："你们这里有卖桂花肉吗？"服务员回答没有。男子仰天长叹："看来我是尽不了孝心了！我老母亲生病已到了最后时刻，她非常想吃桂花肉。我差不多跑遍整个上海，没有一家饭店有供应，很多饭店甚至不晓得啥叫桂花肉！"

正在店里的总经理杨子江听到男子这么一说，非常同情，马上安排他坐下稍等。杨子江学生时代吃过桂花肉，并知道桂花肉怎么做，于是他到厨房，立刻指导厨师做出了一盘桂花肉，让男子赶快送到医院让老母亲品尝。

Longtang Guihuarou (Shanghai-style Fried Pork)

The dish is nostalgically familiar to middle-and-old aged Shanghainese, who often cooked it at home or ate in workplace canteen.

Meat was sliced thickly and dipped in a mixture of egg and wheat flour. After being marinated, it was deep fried, until golden brown, and then stir fried with pepper and salt.

It's crispy outside and tender inside, not stringy or tough at all. It's wonderfully savory, crunchy crackling, and pleasingly addictive.

As its appearance looks like blooming, osmanthus, people called it *guihuarou*, or osmanthus meat.

But the young generation never heard of such a delicacy. Yang Zijiang, general manager of Kong's Garden, felt sorry for it and decided to popularize the dish. Now, his restaurant is the only one in Shanghai supplying *guihuarou*.

提起桂花肉，中年以上的上海人是不会陌生的。那时单位食堂常有供应，家里也经常制作。取五花肉切成略厚的大片，鸡蛋与小麦粉调成蛋糊，将肉片上浆后，炸至金黄，盛起装盆，跟椒盐上桌。桂花肉口感外脆里嫩，鲜香扑鼻。入口后听到咔嚓一声，香气即在口鼻间盘绕，细嚼之下，无筋无渣，吃了还想吃。

这道在老上海人记忆中挥之不去的家常菜，对于今天的年轻人而言，几乎已是闻所未闻的"传奇"。经历此事后，杨子江觉得应该请桂花肉"重出江湖"，于是孔家花园酒家成了目前上海唯一供应桂花肉的饭店。

吃货说：

桂花肉是上海人永远的情结，是上海这座城市的烟火气息，与弄堂生活一样永存于市民的记忆深处。外脆里嫩的五花肉，金黄微焦的葱花香，是艰难时世的亲情与友情的承载。桂花肉不是一般的肉，而是一段温暖人心的市民生活史！

霉干菜焐肉

"绍兴霉干菜"驰名海内外,色泽乌黑,香气浓郁,味美质嫩,久贮不坏,是一种可常年食用的食材。

绍兴还有一道全世界闻名的佳肴:霉干菜焐肉。很多人把它和广东的梅菜扣肉认为是同一道菜,其实大大不然。广东梅菜是用当地一种变种芥菜,而绍兴霉干菜用的是大叶芥、尚未抽苔的白菜和油菜等进行加工而成的。

Meigancai Wurou (Pork Stewed with Preserved Vegetables)

Meigancai, dark preserved cabbage or mustard greens, is widely used by Shaoxing people as side dish or ingredient. *Meigancai wurou*, pork stewed with *meigancai*, is one of Shaoxing's best-known traditional dishes.

The late Premier Zhou Enlai was Shaoxing native, and he loved it.

Once, Zhou dined with Marshal He Long in Louwailou, a time-honored Hangzhou restaurant, and he ordered *meigancai wurou*.

In early 1970s, Zhou treated then US President Nixon with it.

人物简介

周恩来(1898年—1976年),字翔宇,原籍绍兴,生于淮安。中国共产党的主要领导人,中国人民解放军的主要创建人与领导人。自1949年至1976年,担任国务院总理、中共中央副主席等职。

第卅一篇

霉干菜焐肉

烧这道菜,先用上好的带骨五花猪肉,经酱油、白糖等调匀,与浸泡后的霉干菜一层一层夹叠在大海碗中,入大饭锅中蒸。蒸到碗中肉烂菜酥,不分彼此。入口的感觉糯而不腻,吸足了油脂的霉干菜尤其美味。当地民谣:"乌干菜,白米饭,神仙闻到要下凡",可见霉干菜焐肉有多大的魅力!

周恩来对这道浙江土菜情有独钟,有一次,他与贺龙元帅在杭州楼外楼饭店用餐,特意点了霉干菜焐肉。周恩来颇为自豪地对贺龙元帅说:"你尝尝我家乡的土菜,比你喜欢的酸辣汤如何?"贺龙元帅仔细品尝之后连声叫好。七十年代初,周恩来还曾用这道菜宴请了美国总统尼克松。

吃货说:

矮穷丑的绍兴霉干菜,邂逅了高富帅的两头乌猪肉,冲破世俗的偏见,谈了一场如痴如醉的爱情,联手打造了一件美妙的经典杰作——霉干菜焐肉。它从农家的土灶头来到大上海,变的是形式,不变的是美味。

第卅一篇

周恩来

霉干菜焐肉

四星望月

　　位于江西中部的兴国县是大革命时期著名的模范县，一道由毛泽东定名的红色经典菜"四星望月"便源自那里。建国后毛泽东多次巡察到江西，都要品尝此菜。

　　1929年毛泽东率领红四军从井冈山突围，转战赣南，到达兴国县时已经筋疲力尽。兴国县委的负责人将毛泽东请入当地一家小饭馆"打牙祭"，请他吃兴国的一道客家菜："竹笼粉蒸肉"。

Sixing Wangyue (Steamed Pork in Bamboo Baskets)

　　In 1929, revolutionary leader Mao Zedong and his 4th regiment, under the Red Army, the early name of the People's Liberation Army, were besieged by Kuomingtang troops in Jinggang Mountain in Jiangxi Province. Mao led his soldiers to break through the encirclement and ran southward.

　　They were entirely exhausted when they arrived Xingguo County of Jiangxi prowne. Local Party leaders Hu Can and Chen Qihan treated Mao a specialty, steamed pork in bamboo baskets. Four side dishes were served together.

人物简介

　　毛泽东(1893年—1976年)，字润之，湖南湘潭人。马克思主义者，中国无产阶级革命家、战略家、理论家，中国共产党、中国人民解放军、中华人民共和国的主要缔造者，中国各族人民的领袖。毛泽东思想的主要创立者。

四星望月

菜很快上桌了,毛泽东见桌上中间一个圆竹笼里盛着肉片,周围配四碟小菜,形式新颖,口感也佳,便说:"一个圆竹笼像月亮,四个碟子像星星,这跟工农商学盼望我们红军建立根据地一样,不如将此菜叫作'四星望月'吧。"于是这道传统的客家菜,就此载入了中国名菜谱。

此菜在当地还常用鲜鱼来烹制,这里就介绍以猪肉为食材的制法。新鲜五花肉切片,抹上辣糊和米粉待用,先在竹蒸笼里垫上荔芋厚片,蒸熟后再铺上肉片,用旺火急蒸数分钟便可起锅上桌。揭开笼盖,浓烈的鲜香爽辣味扑鼻而来。围边的卤豆腐、酱芋头、酸豆角、腌黄瓜,虽然食材简单,也堪称下酒妙品。

"The round bamboo basket looks like a full moon and four plates surrounding it look like stars. What about to call it *sixing wangyue*, "Mao said. Literally, it means four starts surrounding moon.
Mao made the local dish well-known around China.

吃货说:

　　这道菜中有革命者的豪情,也有江西菜的基因,有鲜香咸辣的滋味,也有星星与月亮的凝望。

第卅二篇

芙蓉鱼排

　　湘菜特级烹饪大师石荫祥，是毛泽东最欣赏的厨师。他制作的一道菜由毛泽东亲自命名，并成为毛泽东最爱吃的菜肴之一。

　　1959年毛泽东回湖南家乡，入住蓉园宾馆后由石荫祥担任主厨。石大厨知道毛泽东爱吃鱼又特别怕鱼刺，于是参照西菜的做法，将鱼剁成细茸涂在面包片上，裹上蛋清后入锅油炸。

Furong Yupai (Deep-fried Fish Cutlet)

　　Hunan cuisine master Shi Yinxiang is Mao's favorite chef, and Shi cooked for him after he returned to Hunan in 1959.

　　Mao loved eating fish but fish bones really worried him. Shi especially invented a deboned fish dish for him, taking western fish cutlet as reference.

　　Shi spread minced fish on bread, dipped in egg white and deep fried it.

　　Mao was pleased with golden, crunchy yet tender cutlet.

　　Initially, Shi called it fish clipped in bread. But Mao didn't agree with such a direct name. He named it after Furong, alias for Hunan.

芙蓉鱼排

上菜后，只见一块块面包片色泽金黄，十分诱人。毛泽东一尝，外脆里嫩，又没有骨刺，鱼的鲜香扑鼻而来，便问石大厨：这菜叫什么名字啊？石荫祥说：就叫面包夹鱼。毛泽东摆摆手说："这个不行，名字太俗，我来帮你改一下，芙蓉是湖南的雅名，就叫芙蓉鱼片吧。"一道新创湘菜就此诞生。

石荫祥晚年总结自己烹饪经验时说："一道好菜一定要好吃，好看，名字好听。"这道"芙蓉鱼排"便是经典了。

> 吃货说：
> 松软的面包衬底，托起如雪的鱼茸，经过温油两次翻炸，成就了外脆里酥的湘江名菜，蘸取多种酱料，也许能品味出风云激荡的往事。

后记

  2015年春节前，我被央视请到北京，与著名节目主持人阿丘做了一次访谈。后来这个节目以《上海故事烩》这样一个极具网络时代特色的名称在CCTV《见证品牌》栏目播出。访谈中阿丘问我：你怎么会想起做一菜一故事？于是我讲了一段很有点"黑色幽默"的往事。

  2001年，有朋友告诉我，开在徐家汇孔祥熙别墅副楼里的小饭店要转让。我喜欢美食，而且一直希望有机会"制造"一些自己喜欢吃的东西与朋友们分享，于是盘下这家饭店，就地命名为"孔家花园"开始了"玩票生涯"。

  没想到小店开张才几天，香港电视台TVB8突然来访，他们正在拍摄《吃遍大中华》，不拍大店，专拍风情小店。导演说：希望拍些有故事的菜肴，并与我约定三天后再来。我想这里曾是孔祥熙的别墅，而孔祥熙与孔夫子同门，于是立时想到一道传说与孔夫子有关的古代名菜："鱼腹藏羊"。俗话说"鱼羊为鲜"，便与此菜有关。我把这个故事讲给香港导演听，导演很开心，决定就拍这道菜。那么这道古董名菜到底怎么做法呢？我与大厨反复折腾，第三天总算交出"作业"。开拍的那时，只见美女主持人讲得眉飞色舞，还学着孔老夫子的样子，文绉绉地吃了口鱼，拉长了声调说道："如此鲜美，实乃阴阳相合，从此鱼羊为鲜矣。"其实，她面前的那条鱼半生半焦，实在无法下口！

  真正研制成功"鱼腹藏羊"这道极具历史文化内涵的名菜，是在电视片的拍摄之后。我与大厨花了很长时间琢磨，不用任何人工鲜味剂，并且保持"烤"的特色，把"鱼腹藏羊"这道菜做成了真正的"鱼羊为鲜"，最终成为了饭店的招牌菜。一时间，"尝鲜字来历，到孔家花园"变成了当时报纸上的经典广告语。孔家花园顾客盈门，由此开始。

2006年，饭店从别墅副楼搬进了别墅主楼，广告语也变成了"一座有故事的花园，一家有好菜的饭店"。挖掘菜肴背后的故事，提升饮食文化的层次，成为了饭店的社会责任和核心竞争力。这些"故事菜"，都是我从各种书中"淘"出来的。故事生动有趣，有教育意义，有中国特色，这是选择的标准。更重要的是，这道菜必须是独创而不能流于平常，必须做得好吃，能够让顾客在美味的享受中获得物质与精神的双重超值享受。

做餐饮做到这个份上，我觉得越来越有意思了！我也从"玩票"转业为"入行"。一位老朋友说，只有你这种混过报社再混入餐饮界的"杂种"，才会玩得出一菜一故事！可能正因为我们的餐饮界"杂种"太少，才让我有了如此机会！十五年"玩"出了这本小书。

这本小书里精选的三十三道菜肴，每一道菜的食材配比、营养成分、烹饪火候、摆盘形式，都是经过无数次试验才得以确定下来，每一道菜肴如何从历史传说回归到现实餐桌，本身也是一个故事。这是我与大厨们、以及不断对菜肴提出意见的新老顾客们集体智慧的结晶！

入选央视节目除了上述的《上海故事烩》之外，还有《领航》栏目里的《聆听夜上海 品味民国菜》。上海近五万家饭店，各擅胜场，各有绝活，何以唯独我店入选央视节目，而且两次，受宠若惊之余我曾向央视导演请教个中原因。央视导演说：如按饭店菜肴的特色来选择，众口铄金，没有一个让大家心悦诚服的标准，很难操作。而以文化结合菜肴，以菜肴弘扬中国餐饮文化，这个特色至今为止，只有你店。

那么入选央视节目的原因，应该也是入选上海书店出版社选题的原因。能以一家饭店菜谱上的菜创作出一本连环画，可能又是餐饮界至今的唯一"传奇"。为此，我深深感恩源远流长的中华餐饮文化，感恩多年来一直鼓励我，并且关怀着"一菜一故事"一个个有如娃娃，从诞生到长大的各界朋友们老师们，感恩十多年来欣赏品味这些菜肴的顾客朋友们！

如今，因为十年租约期满，第一家"孔家花园饭店"已离开"花园"。新的门店在杨浦五角场商圈、浦东八佰伴商圈、淮海中路商圈乃至莘闵别墅区相继开张，特色连锁餐厅也正在积极筹划中。根据诸多朋友与顾客的意见，离开孔家别墅的饭店再姓"孔"似不合适，因此决定将饭店更名为"流金年代"。不容置疑的是，我们共同打造的"一菜一故事"将如灿烂黄金，从中华民族辉煌历史的长河中，源源不断地流向大众的餐桌！

这本小书中的所有菜肴，在我们各家门店都能品尝得到。希望诸位朋友相约三五知己，光临小店，一边品尝美味佳肴，一边穿越时空，与画册中的英雄美人同坐花下，更进美酒，这才叫"酒不醉人人自醉"，"风景这边独好"！

杨子江

2016.7.18.

Postscript

Before last year's Spring Festival, I got interviewed by popular TV anchor A Qiu for stories behind food in Kongjia Huayuan, made famous by our distinctive menu that features dishes in connection with past-time celebrities.

In fact, the idea to make stories the heart and soul of my restaurant didn't initially occurred to me though it's later proved pretty profound. Yes, it's an accidental business-success story.

In 2001, I bought a restaurant opened in the former residence of Kung Hsiang-hsi. I started the business more for the sake of fun, hoping it a rendezvous for my friends to enjoy culinary delights. In honor of Kung, I called it Kongjia Huayuan, or Kung's villa.

It was only a few days after I plunged into the catering world when an unexpected chance to make Kongjia Huayuan known came my way.

At the time, Hong Kong-based TVB8 channel was shooting Chinese distinctive restaurants for its new program aimed at presenting Chinese culinary culture. Our address interested the director, who wanted me to discover stories behind food as to match the "narrative" tone of the historical villa.

The legendary *yufu cangyang*, roasted fish stuffed with mutton, immediately came to my mind. It's one of the favorite dishes of Confucius, who's said the ancestor of Kung. Confucius thought the combination of fish and mutton makes the ultimate umami, which caused the evolution of Chinese character of 鲜 (umami), combining 鱼(Chinese character of fish) with 羊 (Chinese character of mutton).

The director was glad to hear that, and gave me three days to prepare.

Chef and I made a good-looking but terrible-tasting version.

On the shooting day, the TV hostess imitated Confucius's way of talking, saying: "Soooo umami!" In front of the camera, it seemed a super-savory deliciousness, but I knew it was just a failure. The fish was overcooked while the mutton was still raw.

When we really made it was some time later. All ingredients were natural, without artificial flavoring. And it became our signature dish. Diners came from everywhere to have a taste of history.

Over years, I red a lot to explore what goes on behind the scenes, and chefs turned recipes from books to tables in real life.

In 2006, Kongjia Huayuan moved from the villa's annex to its main body when it has been a hot dining spot in the city, at which point I decided to take it seriously as my life-long career.

One of my old friends called me "half-blood" catering man as I used to work for newspaper.

"Only people like you can play fusion well," he said.

In this book, I illustrated 33 dishes served in my restaurant. Every one of them is the brainchild of my chefs through numerous trials.

The CCTV director told me that the idea to infuse culture to food is so unique, which may give reason to why Kongjia Huayuan was selected for shooting from about 50,000 restaurants in Shanghai.

Maybe it's also why Shanghai Bookstore Publishing House published a comic book inspired by my menu, which as far as I know, is unprecedented.

I have to thank for the great Chinese culinary culture, and also my customers and my friends for their care and advice.

As the lease was up, Kongjia Huayuan moved out from Kung's villa. And the restaurant name has thus been changed to the Golden Age, while the menu remains the same.

Now, you can find all 33 dishes in our four new outlets, which were decorated to connote sort of old-time chic, in Pudong New Area, and districts of Yangpu, Huangpu and Minhang.

Bon appetite!

July 18, 2016, Yang Zijiang

图书在版编目（CIP）数据

一菜一故事 / 杨子江主编； 罗希贤，罗一绘画；沈嘉禄编文 -- 上海：上海书店出版社，2016.8
ISBN 978-7-5458-1331-9

I. ①一⋯ II. ①杨⋯ ②罗⋯ ③沈⋯ ④罗⋯ III. ①饮食—文化—中国—通俗读物 IV. ① TS971-49

中国版本图书馆 CIP 数据核字（2016）第 176654 号

一菜一故事

主　编	杨子江	责任编辑	杨柏伟　邢侠
绘　画	罗希贤　罗一	英文翻译	李倩　高层
编　文	沈嘉禄	装帧设计	罗一
书　法	陈志宏	技术编辑	丁多
篆　刻	杨忠明	美术编辑	汪昊

出　版　上海世纪出版股份有限公司上海书店出版社
发　行　上海世纪出版股份有限公司发行中心
地　址　200001　上海福建中路 193 号
　　　　　www.ewen.co
印　刷　上海展强印刷有限公司
开　本　889×1194　1/24
印　张　6
版　次　2016 年 8 月第一版
印　次　2016 年 8 月第一次印刷
书　号　ISBN 978-7-5458-1331-9/TS.6
定　价　25.00 元